Using an
EXPLOSIVE SUBSTANCE
to
GENERATE ELECTRICITY

Robert J. Richey

authorHOUSE®

AuthorHouse™
1663 Liberty Drive
Bloomington, IN 47403
www.authorhouse.com
Phone: 1-800-839-8640

Published by AuthorHouse   11/7/12

ISBN: 978-1-4772-8693-7 (sc)
ISBN: 978-1-4772-8692-0 (e)

Library of Congress Control Number: 2012920716

Any people depicted in stock imagery provided by Thinkstock are models, and such images are being used for illustrative purposes only. Certain stock imagery © Thinkstock.

This book is printed on acid-free paper.

Because of the dynamic nature of the Internet, any web addresses or links contained in this book may have changed since publication and may no longer be valid. The views expressed in this work are solely those of the author and do not necessarily reflect the views of the publisher, and the publisher hereby disclaims any responsibility for them.

CONTENTS

OTHER BOOKS BY THIS AUTHOR

1. Take Time to Smell the Roses Book of Poetry
   - ISBN # 1-4033-4472-8 (Paperback)
   - ISBN # 1-4033-8753-2 (Hardcover)

2. The Golden Knight Book of Chess: The Art of Sacrifice
   - ISBN # 1-4208-6573-0 (Paperback: Text in Black & White)

3. Life on the Diamond Bar Ranch: A Tale of the West
   - ISBN # 978-1-4259-6451-1 (Paperback 6" x 9")

4. Robinhood of the Underworld: Dominic Capizzi
   - ISBN # 978-1-4343-1949-4 (Paperback)

5. Life on the Diamond Bar Ranch: A Tale of the West

- ISBN # 1-4259-2982-6 (sc)(Text & Pictures in color; 8 ½" x 11")

6. The Unlikely Hero: A Tale of the Sea
   - ISBN # 978-1-4343-9892-5 (Paperback)

7. The Unlikely Hero: A Tale of the Sea
   - ISBN # 978-1-4343-9891-8 (Hardcover)

8. Mutiny in the United States Navy in World War II: A True Story
   - ISBN # 978-1-4389-6047-0 (Paperback)
   - ISBN # 978-1-4389-6048-7 (Hardcover)

9. Destroyer Squadron 12 in the Solomon's Campaign: The Tip of the Lance
   - ISBN # 978-1-4490-5268-3 (sc)

10. Blitz Chess Puzzles; The Art of Sacrifice
    - ISBN: #978-1-4520-4798-0 (e)
    - ISBN: #978-1-4520-4797-3 (sc)

11. Key Chess Puzzles: Sacrificial Chess
    - ISBN: #978-1-4520-8753-5 (sc)

- ISBN: #978-4520-8754-2 (e)

12. My Brother Glenn a Prisoner of the Gestapo in World War II: German Secret Police.
    - ISBN: #978-4567-6687-0 (sc)
    - ISBN: #978-1-6688-7 (e)

13. Rescue of the Helena Survivors in the Solomon Island Campaign: A Tale of Incredible Valor
    - ISBN: #978-1-4670-3857-7(sc)
    - ISBN: #978-1-4670-3855-3(hc)
    - ISBN: #978-1-4685-0199-5(e)

14. Bringing Braille into the Computer Age: Carrying on the Torch
    - ISBN: (978-1-4685-2) (sc)
    - ISBN: (978-1-4685-5) (e)

THE DILEMMA OF THE DAY

*One of the World's Dilemmas of the Day
is the increasing Over Population.*

File:Population curve.svg
From Wikipedia, the free encyclopedia

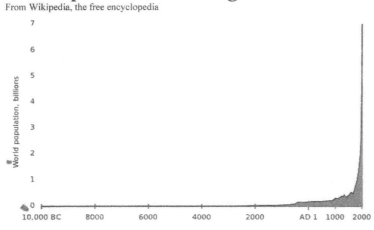

It has been said, "that a picture is worth a thousand words."
This graph demonstrates dramatically how explosively the
World Population has increased in the past thousand years.
In fact, in the past few decades it has virtually gone off of

the chart. It has been reported that China has passed laws forbidding couples from having more than one child. But that is a leaky boat, as the same couples can immigrate to another country that does not have such laws. In this new place they can have as many children as they desire to have.

World Population Estimate

Year	Billions	Difference
1800	1	
1927	2	127 years
1960	3	33
1974	4	14
1987	5	13
1999	6	12
2011	7	12

Estimate by the United States Fund estimated in 10/31/2011

The above chart demonstrates that the time between successive increases of an additional billion of new persons grows shorter and shorter.

For Example: Between the year 1800 and the year 1927 the World's population increased from 1 billion to 2 billion. This increase occurred in just 127 years.

But the next increase from 1927 to just 1960 a third billion increase took place raising the total to 3 billion. This increase only took 33 years.

From 1960 to 1974 a fourth increase to 4 billion occurred. This only spanned a mere 14 years. It is very evident that the progressive was accelerating.

From 1974 to 1987, a mere 13 years, the increase to 5 billion occurred.

From 1987 to 1999, in a mere 12 years the increase to 6 billion occurred.

And finally from 1999 to 2011, again in a span of only 12 years the increase to 7 billion occurred.

It is painfully obvious that if this increase continues, or in fact accelerates, Big Trouble Lurks in the Wings.

If, at the present the Countries of the World are struggling to feed their people, what will it be like in the future?

Over Population inevitably leads to starvation, disease and social unrest on a grand scale.

The World Leaders have been only too aware of the problem but there seems to be that there isn't any solution.

Like a herd of buffalo fleeing from the flames of a wind driven grass fire, they figuratively refuse to find a solution to the problem before hurtling over the cliff. The Cliff may not be that far off!!

A SECOND DILEMMA
OF THE DAY

This Dilemma consists of the continuing need for the Energy. This is involved largely with the generation of Electrical Power.

Fossil fuels have been and will continue to be the primary source of Energy. In the year 2007 it was estimated that the primary sources consisted of petroleum 36.0%, coal 27.4%, natural gas 23.0%. These figures total up to 86.4% for fossil fuels as the primary energy source world wide. Other sources consist of hydroelectric 6.53%, Nuclear 8.5%, while about 0.9% of the total energy needs consist of geothermal, solar, wind, burning of wood and waste. It was estimated that the World energy consumption is growing at 2.3%.

The downside to the use of fossil fuels is the massive production of CO_2 estimated at 21.3 billion tones per year. It is estimated that natural processes can only remove 50% of that which is being produced. This discrepancy leads to an increase of 10.65 billion tones of atmospheric carbon dioxide per year. It is believed this is causing warming of the atmospheric that is resulting in rising ocean levels and disruption of normal weather patterns. This disruption leads to massive flooding and the attendant damages. Also, on the other hand it is believed to be contributing to droughts which are especially damaging in the farming areas.

Only certain Countries have large petroleum deposits and most other countries must import large amounts of fossil fuels on a daily basis.

Prior to the latter part of the 18th century windmills and waterwheels provided significant amounts of energy. Steam engines were fueled, first by using coal and later with petroleum products.

A further expansion for the need of petroleum products came about with the invention of the combustion engine. The locomotives and aircraft also contributed to a further need for energy.

The most troubling problem associated with the current method of providing energy with use of fossil fuels is its potential shortage.

Currently the primary sources of energy production are provided for by 79.6% from coal, oil and natural gas.

The proved reserves of the various fossil fuels in 2005-2006:

- For Coal an estimated 997,748 million short tones.

- For Oil an estimated 1,119 billion barrels.

- For Natural gas an estimated 6,183 – 6,381 trillion cubic feet.

Proven reserves consist of the total of fossil fuels that can be brought up from the ground. Certain substances are not recoverable.

It is further estimated that the number of years in which energy that will be required can be produce from the current proven reserves is estimated to be:

- For Coal a total of 148 years.

- For Oil a total of 43 years.

- For Natural Gas a total of 61 years.

Disturbingly for the United States is that even though the United States only accounts for less than 5% of the world's population it never the less uses more than a quarter of the world's fossil fuels.

The combustion of fossil fuels generates air pollutants, such as carbon dioxide, nitrogen oxides, sulfur dioxide, and other volatile organic compounds and also toxic heavy metals.

Although it is true the Electric Car produces less pollution while in use, the electricity to power it has to be produced somewhere. Only the source of the production of pollution has been moved.

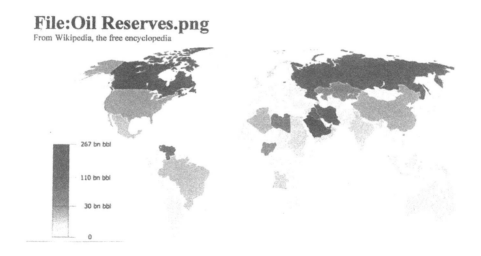

File:Oil Reserves.png
From Wikipedia, the free encyclopedia

World Proven Crude Oil Reserves

Middle East	64.5%	694,605,500,000	Barrels
N/S America	11.5%	124,325,000,000	Barrels
Africa	8.9%	95,461,500,000	Barrels
E. Europe	6.2%	67,159,700.000	Barrels
N. America	2.8%	30,491,000,000	Barrels
W. Europe	2.0%	21,065,600,000	Barrels

Coal Reserves in the United States

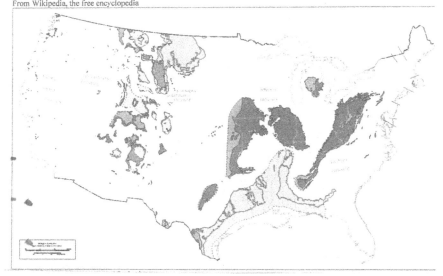

The Estimated Proven Coal Reserves in the United States is as follows:

- Bituminous & Lignite 111,338 million tones.

- Sub-Bituminous & Lignite 135,305 million tonnes.

COAL PROVEN RESERVES WORLDWIDE

Rank	Country	Bituminous & Lignite	Sub-Bituminous & Lignite	Total
1.	**USA**	**111,338**	**135,305**	**246,643**
2.	Pakistan	0	175,300	175,300
3.	Russia	49,088	107,922	157,010
4.	China	62,200	52,300	114,500
5.	India	90,085	2,360	92,445
6.	Australia	38,600	39,900	78,500
7.	South Africa	8,750	0	48,750

All other Countries vary from 0 to 16,274

All Measures in ***Million Tonnes***

PROVEN NATURAL GAS RESERVES IN THE UNITED STATES

File:Shaleusa2.jpg
From Wikipedia, the free encyclopedia

United States Shale Gas Plays

Proven Natural Gas Preserves World Wide

Russia	42,300
Iran	24,021
Qatar	23,191
Saudi Arabia	6,010
U.A.E.	5,454
USA	*4,711*
Nigeria	4,497
Algeria	4,070
Venezuela	*3,734*
Iraq	2,798

All Measurements in Million Tonnes of Oil Equivalents

VENEZUELA PROVEN OIL RESERVE MAP

File:Orinoco USGS.jpg

From Wikipedia, the free encyclopedia

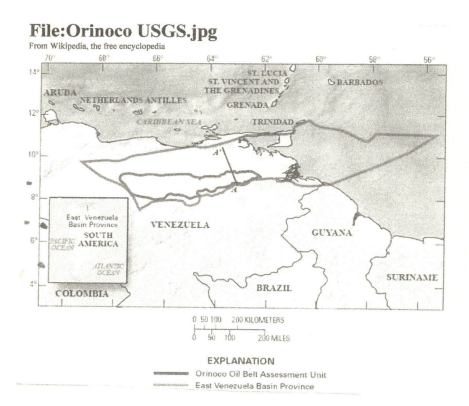

EXPLANATION

━━━━ Orinoco Oil Belt Assessment Unit

▬▬▬ East Venezuela Basin Province

It is presently claimed that Venezuela has the largest Proven Oil Supplies in the World.

According to President Hugo Chavez and the Venezuela government these Reserves reach 297 billion barrels, surpassing Saudi Arabia that has been the previous long time world leader as having the most Proven Oil Reserves.

FACTS

The United States is _Fifth_ in the World in Proven Oil Reserves.

The United States is _First_ in the World in Proven Coal Reserves. but, the government is advocating the reduction of burning Coal.

The United States is _Sixth_ in the World in Natural Gas Proven Reserves.

The United States Consumes more Energy than any of the other Countries in the World.

It is evident that this Country cannot drill its way to oil independence. At present It cannot survive without coal as a dominant fuel.

A more practical method of producing electricity must be found, and soon.

CONCLUSION

All of the above information leads inevitably to the obvious conclusion, The time to find a different ,more permanent method of generating Electricity must be found, and soon.

The time for procrastination is over. The hour is late. It has been said ostriches bury their heads in the sand pretending that if they cannot see danger it does not exist. But we are not ostriches and the danger is real and it does exist. The day is not far off when the wheels of Industry will grind to a halt.

In the pages of this book is a copy of an Invention which has recently been granted a Patent by the US Patent Agency.

It is the belief by this writer, and Inventor, that this particular Invention is capable of producing Electricity in significant amounts. It is also believed that the Source of Power proposed is inexhaustible and not subject to the vagaries of Nature.

THE PATENT
FOR USING AN EXPLOSIVE
SUBSTANCE FOR GENERATING
ELECTRICITY

Note the following:

A Copy of the Invention can be obtained off of the Internet on Google.

Step 1.   Boot up the Computer

Step 2.   Call up the Google Screen

Step 3.   In the Search Block on the Google Screen type in "Google Patent Search" & Click on the Search Icon

Step 4.   On the New Pop-up Screen type in the number 8,215,111 and click on the Search Icon

Step 5.   Scroll down to the Website for the Patent and click

(12) **United States Patent**
    Richey

(10) **Patent No.:** US 8,215,111 B1
(45) **Date of Patent:** Jul. 10, 2012

(54) **ELECTRICAL GENERATION FROM EXPLOSIVES**

(76) Inventor: **Robert J. Richey**, Campbell, CA (US)

(*) Notice: Subject to any disclaimer, the term of this patent is extended or adjusted under 35 U.S.C. 154(b) by 502 days.

(21) Appl. No.: **12/386,822**

(22) Filed: **Apr. 23, 2009**

Related U.S. Application Data

(60) Provisional application No. 61/100,915, filed on Sep. 29, 2008.

(51) **Int. Cl.**
    F01B 29/00   (2006.01)
    F16D 31/00   (2006.01)
    F16D 39/00   (2006.01)
(52) **U.S. Cl.** 60/512; 60/325; 60/415
(58) **Field of Classification Search** 60/369, 60/484, 516, 632, 634, 675, 914
    See application file for complete search history.

(56) **References Cited**

U.S. PATENT DOCUMENTS

178,925 A * 6/1876 Hardy 60/634
3,611,723 A * 10/1971 Theis 60/327

3,648,458 A	*	3/1972	McAlister 60/415
3,650,572 A		3/1972	McClure
4,301,774 A		11/1981	Williams
5,551,237 A	*	9/1996	Johnson 60/641.8
5,647,734 A	*	7/1997	Millcron 417/380
5,713,202 A	*	2/1998	Johnson 60/325
5,865,086 A	*	2/1999	Petichakis P. .... 91/4 R
6,182,615 B1	*	2/2001	Kershaw 123/19
6,739,131 B1	*	5/2004	Kershaw 60/512
7,531,908 B2	*	5/2009	Fries et al. .... 290/1 R

FOREIGN PATENT DOCUMENTS

WO   WO 01/31197 A1   3/2001

* cited by examiner

Primary Examiner — Thomas Denion
Assistant Examiner — Christopher Jetton

(57) **ABSTRACT**

An electrical generator converts the high blast pressures of explosives into useful electricity by capturing the explosive gases and using the high gas pressures to alternately push water hydraulically between two tanks and through water turbines connected to DC electric generators. Water expelled through a water turbine from one tank is used to fill the other tank. Batteries can be used to store the electrical energy generated, and inverters followed by transformers convert the DC electric from the turbine-generators to 110-VAC, 220-VAC, and 440-VAC. A microcomputer controller connected to various sensors and solenoid valves coordinates the timing and routing of the detonation of explosives, tank pressures, venting, valving, and load control.

9 Claims, 4 Drawing Sheets

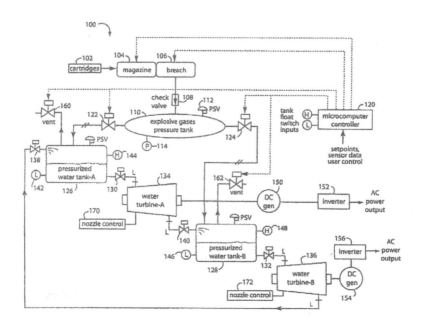

Fig. 1

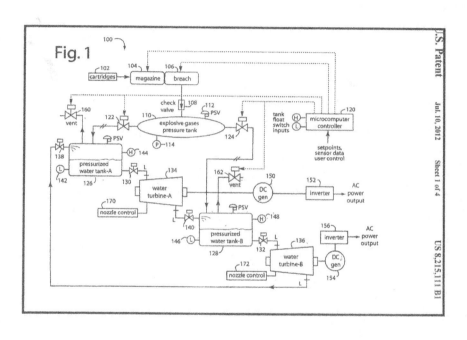

Fig. 2

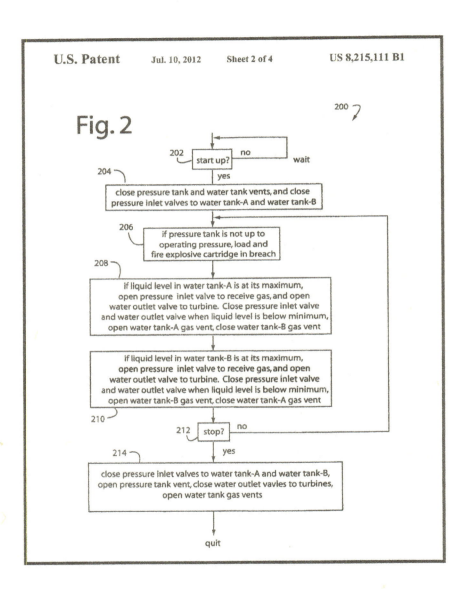

200

202 start up?   no   wait

yes

204 close pressure tank and water tank vents, and close pressure inlet valves to water tank-A and water tank-B

206 if pressure tank is not up to operating pressure, load and fire explosive cartridge in breach

208 if liquid level in water tank-A is at its maximum, open pressure inlet valve to receive gas, and open water outlet valve to turbine. Close pressure inlet valve and water outlet valve when liquid level is below minimum, open water tank-A gas vent, close water tank-B gas vent

if liquid level in water tank-B is at its maximum, open pressure inlet valve to receive gas, and open water outlet valve to turbine. Close pressure inlet valve and water outlet valve when liquid level is below minimum, open water tank-B gas vent, close water tank-A gas vent

210

212 stop?   no

yes

214 close pressure inlet valves to water tank-A and water tank-B, open pressure tank vent, close water outlet vavles to turbines, open water tank gas vents

quit

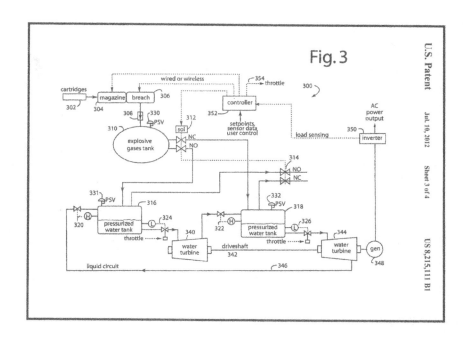

Fig. 3

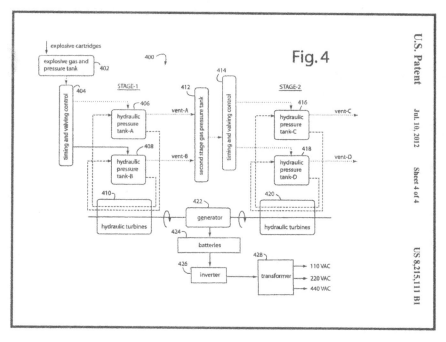

Fig. 4

1

ELECTRICAL GENERATION FROM EXPLOSIVES

RELATED APPLICATIONS

This Application claims benefit of U.S. Provisional Patent Application, A UNIQUE METHOD OF GENERATING ELECTRICITY USING EXPLOSIVE SUBSTANCES AS A POWER SOURCE, Ser. No. 61/100,915, filed Sep. 29, 2008.

BACKGROUND OF THE INVENTION

1. Field of the Invention

The present invention relates to electrical power generation, and in particular to devices and methods for converting the gas pressure generated by explosives into electricity.

2. Description of the Prior Art

Useful electrical energy does not exist in nature and it must be converted from other available energy forms such as gasoline, diesel, coal, natural gas, geothermal, steam, hydro, solar, etc. Some of these energy sources are expensive, some are highly pollutant, some are difficult to convert efficiently, and some are not very portable as is needed in vehicles.

Explosives like gunpowder; Cordite, Ballistite, and Poudre-B smokeless powders; trinitrotoluene (TNT), Dynamite; nitroglycerin; Tovex and other water gel explosives; etc., release a lot of energy in a very rapid pulse. Explosives usually have less potential energy than petroleum fuels, but their high rate of energy release produces large blast pressures. TNT has a detonation velocity of 6,940 m/s compared to 1,680 m/s for the detonation of a pentane-air mixture, and the 0.34-m/s stoichiometric flame speed of gasoline combustion in air. Explosives are classified as "low" or "high" explosives according to their rates of decomposition. Gunpowder is a low explosive, while TNT is a high explosive. Low explosives burn rapidly or deflagrate, while high explosives detonate.

The energy released includes high levels of heat, light, and gas pressure. These are all quickly dissipated if not captured or otherwise contained. For example, at 15° C. the volume of gas produced by the explosive decomposition of one mole of nitroglycerin, becomes, $V=(23.64 \text{ liter/mol})(7.25 \text{ mol})=171.4$ liters. The molar volume of an ideal gas at 15° C. is about 23.64 liters. The potential of an explosive is the total work that can be performed by the gas generated by the explosion. If uncontained, it expands adiabatically from its original volume until its pressure is reduced to atmospheric pressure and its temperature to ambient.

In the nitroglycerin reaction, $C_2H_5(NO_2)3 \rightarrow 3CO_2 + 2.5H_2O + 1.5N_2 + 0.25O_2$, the products are carbon dioxide, water, nitrogen, oxygen, and heat. Therefore, a relatively small solid or liquid volume is converted into a very large volume of relatively benign gases. Nitroglycerin explosions are relatively clean, compared to TNT which is poisonous and produces a lot of carbon soot in its reaction.

Firearms and artillery use the gas pressure generated by the detonation of smokeless powder to accelerate bullets and projectiles to very high muzzle velocities on the order of 2,000+ feet per second. Sticks of explosives are detonated in holes drilled into geologic deposits to fracture the ores and make removing the material as easy as scooping up the pieces.

What is needed is a device and method to convert explosive energy into a more useful form of electrical energy as used in homes and industry.

SUMMARY OF THE INVENTION

Briefly, an electrical generator embodiment of the present invention converts the high blast pressures of explosives into

2

useful electricity by capturing the explosive gases and using the high gas pressures to alternately push water hydraulically between two tanks and through water turbines connected to DC electric generators. Water expelled through a water turbine from one tank is used to fill the other tank. Batteries can be used to store the electrical energy generated, and inverters followed by transformers convert the DC electric from the turbine-generators to 110-VAC, 220-VAC, and 440-VAC. A microcomputer controller connected to various sensors and solenoid valves coordinates the detonation of the explosives, tank pressures, venting, valving, and load control.

These and other objects and advantages of the present invention will no doubt become obvious to those of ordinary skill in the art after having read the following detailed description of the preferred embodiments which are illustrated in the various drawing figures.

IN THE DRAWINGS

FIG. 1 is a simplified functional block diagram of a single-stage electrical generator embodiment of the present invention that cycles pressurized water between two tanks and through two sets of water turbines;

FIG. 2 is a flowchart diagram of an electrical generator method embodiment of the present invention to cycle pressurized water between two tanks and two water turbines, as in FIG. 1;

FIG. 3 is a functional block diagram of a single-stage electrical generator embodiment of the present invention that cycles pressurized water between two tanks and through two water turbines like that of FIG. 1, but that reduces duplication of the DC generators and inverters, and some of the valving; and

FIG. 4 is a simplified functional block diagram of a two-stage electrical generator embodiment of the present invention that uses explosive gases to pressurize water, and then uses pressurized hydraulics to spin electrical generators with hydraulic motors and turbines.

DETAILED DESCRIPTION OF THE PREFERRED EMBODIMENT

In FIGS. 1-4 and the following text, some of the more conventional and routine elements commonly used with gas and hydraulic valves, pressure tanks, plumbing, and process control systems are not shown or described. For example, inspection ports and drains for water tanks, safety relief valves, check valves, nozzles for turbines, gearboxes and pulleys, wireless interfaces, wiring, etc. The components like these that should be used are engineering choices and are routinely stocked and installed by technicians. The critical and unusual combinations and their interrelationships are described here in detail.

FIG. 1 represents a single-stage electrical generator system embodiment of the present invention, and is referred to herein by the general reference numeral 100. Generator system 100 produces electrical power suitable for homes, businesses, industry, and the utility grid from the explosive energy captured from cartridges 102 loaded in a magazine 104 and fired in a breach 106. Cartridges 102 should include low explosives that burn clean and soot-free, and the chemical reactions should not produce any dangerous gases or byproducts. For example, nitroglycerin reactions only produce carbon dioxide, water, nitrogen, oxygen, and heat. The heat actually helps increase the gas pressures up to operating levels and should not be wasted or exhausted until the maximum in work has been extracted.

3

The heated gaseous explosive products are passed through a check valve 108 to a pressurized-gas tank 110. A pressure safety valve (PSV) 112 provides relief if the internal pressures exceed a safe maximum. A pressure sensor (P) 114 measures the tank pressures for a microcomputer controller 120. In some installations the pressure readings will be reported wirelessly, in others a simple 4-20 milliamp process control loop can be used.

Microcomputer controller 120 coordinates all the timing and valve control needed to operate generator system 100 and keep it safe. It uses readings from pressure sensor (P) 114 to determine when more cartridges 102 need to be loaded in magazine 104 and fired in breach 106, and it controls the actual firing. Microcomputer controller 120 also decides when and which gas pressure inlet valve 122 and 124 should be opened and closed for pressurized water tank-A 126 and pressurized water tank-B 128.

Pressurized water tank-A 126 and pressurized water tank-B 128 are not simultaneously pressurized, the pressure applied to them is alternated by gas pressure inlet valves 122 and 124 under control of microcomputer controller 120. What's important to the timing is the water levels inside the tanks, there are minimum and maximum operating levels that must be respected. Water inside one tank needs to flow out into the other tank through a water turbine, and the water cannot flow if the receiving water tank is pressurized at the same time.

In FIG. 1, an outlet valve-A 130 is opened to pass pressured liquid water (L) to a water turbine-A 134. Similarly, an outlet valve-B 132 is opened to pass pressured liquid water (L) to a water turbine-B 136. The liquid water returns from water turbine-B 136 through an inlet valve-A 138 back to water tank-A 126. Liquid water from water turbine-A 134 passes on through to inlet valve-B 140 to water tank-B 128.

The minimum and maximum operating levels of water that circulate between water tank-A 126 and water tank-B 128 are set by float switches (L, H) 142 and 144 for water tank-A 126, and by float switches (L, H) 146 and 148 for water tank-B 128. These float switches are connected to microcomputer controller 120, and the readings are used to determine when to open and close outlet valve-A 130, outlet valve-B 132, inlet valve-A 138, and inlet valve-B 140. The float switch connection could be done wirelessly, and a local loop could be included to automatically close, for example, water outlet valve-A 130 when minimum level float switch 142 senses low water.

Each water tank-A 126 and water tank-B 128 should be equipped with a water to add make up water, and to drain water completely, e.g., during maintenance.

In FIG. 1, water turbine-A 134 is mechanically connected by a rotating shaft to drive a DC electrical generator 150. The DC electrical power produced could be stored in batteries, and it is converted to AC electrical power by an inverter 152. Similarly, water turbine-B 136 is mechanically connected by a rotating shaft to drive another DC electrical generator 154. The DC electrical power produced could be stored in the same batteries, and it can also be independently converted to AC electrical power by an inverter 156. The voltage outputs of inverters 152 and 156 can be stepped-up or stepped-down by conventional transformers as needed, e.g., to 110-VAC, 220-VAC, and 440-VAC.

Microcomputer controller 120 is connected to sense the electrical loads placed on inverters 152 and 156, and uses the information to control how much pressurized water is needed to be passed through water turbine-A 134 and water turbine-B 136 to keep the overall operation in balance.

4

Once the pressurized gas inside the water tanks has done its job pushing out the water down to its minimum operating level, the residual pressurized gas can be vented out. For water tank-A 126, a vent valve-A 160 is used, and for water tank-B 128, a vent valve-A 162 is used. The residual gas pressures can be high enough to do useful work in a second stage generator. But any back pressure caused by the use of later stages can reduce the efficiency of the earlier stages by reducing the differential pressures between the pressurized tank and the vented one.

In operation, falling water levels inside the water tanks can be used by the minimum-level float switches 142 and 146 to trigger closed the associated water outlet valves 130 and 132. This, in turn can be used to trigger closed the gas pressure inlet valves 122 and 124, and to trigger open the gas pressure vent valves 160 and 162. Similarly, the maximum-level float switches 144 and 148 can be used to trigger closed the water inlet valves 138 and 140.

Pressurized water tank-A 126 and pressurized water tank-B 128 would normally be equipped with various conventional items not shown in FIG. 1. For example, inspection ports, drain valves, pressure gauges, pressure safety valves to release excess pressure, and a water make-up input to replace lost water.

Microcomputer controller 120 can increase and decrease the torque outputs of water turbine-A 134 and water turbine-B 136 by sending modulation controls to nozzle controls 170 and 172. Alternatively, water outlet valve-A 130 and water outlet valve-B 132 could be continuously adjustable, instead of simple fully open, fully closed solenoid types. Such torque modulation would be necessary in some applications to balance the power being generated with the loads applied. In such case, inverters 152 and 156 would also be required to provide load measurements to microcomputer controller 120.

FIG. 2 represents an electrical generator method embodiment of the present invention to cycle pressurized water between two tanks and two water turbines, as in FIG. 1. Such method is referred to herein by the general reference numeral 200. Method 200 is implemented as a computer program in software or firmware executed by a conventional microcomputer, e.g., microcomputer controller 120 (FIG. 1). Data inputs from sensors and switches are digitized for processing, user inputs are used to make process control decisions, and outputs to electro-mechanical solenoids are used to operate gas and hydraulic valves.

Method 200 includes three phases of operation: (1) startup, (2) power generation, and (3) shutdown. During startup, the operational pressures and valve settings must be initialized. During power generation, the gas pressure generated by the explosive cartridges must be switched between the two water tanks according to the respective water levels inside each. The amount of water forced between the water tanks and through the water turbines must be balanced with the electrical loads being placed on the system. During shutdown, the cycling must be stopped and the pressures relieved by opening the various vents.

Specifically, method 200 includes a step 202 for checking to see if the user wants to begin operation. If so, a step 204 closes the pressure tank and water tanks vents, and closes the inlet valves to the water tanks. A step 206 gets the gas pressure in the pressure tank up to operating levels by firing explosive cartridges as needed. A step 208 checks the water level inside water tank-A and if it's at its maximum operating level, a hydraulic cycle can begin. The gas inlet valve-A is opened, the gas vent valve-A is closed, and the water outlet valve-A to the associated turbine-A is opened. The gas pressure let in will push the water out through the outlet valve-A. When the water level reaches minimum, the outlet valve-A is closed. The gas inlet valve-A is closed, and the gas vent valve-A is opened. The water inlet valve-A is opened to receive water from water tank-B.

A step 210 checks the water level inside water tank-B and if it's at its maximum operating level, a hydraulic cycle can begin. The gas inlet valve-B is opened, the water outlet valve-B is closed, and the water outlet valve-B to the associated turbine-B is opened. The gas pressure let in will push the water out through the outlet valve-B. When the water level in water tank-B reaches minimum, the outlet valve-B is closed. The gas inlet valve-B is closed, and the gas vent valve-B is opened. The water inlet valve-B is opened to receive water from water tank-A.

If the user is requesting a stop of operations, a step 212 passes control to a step 214. Otherwise, the process repeats in a loop back to step 206. Step 214 closes the gas inlet pressure valves to water tank-A and tank-B, opens the vents, and closes the water outlet valves to the turbines. Residual gas pressures inside the pressurized tank may be let down if another use cycle is not expected immediately.

FIG. 3 illustrates a single-stage system 300 that eliminates some of the duplication of the major components appearing in FIG. 1. System 300 assumes that when the water level in a water tank is below minimum, e.g., as detected by a low-water float switch, the water outlet valve should be closed. Similarly, when the water level in a water tank is above maximum, e.g., as detected by a high-water float switch, the water inlet valve should be closed. The gas inlet valve to a water tank can only be open if the gas vent is closed. The gas inlet valve to the water tank must be closed if the gas vent is open. The mechanisms implemented in such logic can be built with relay logic, software, IC logic gates, and mechanical interlocks.

System 300 is powered by explosive cartridges 302 that are loaded in a magazine 304 and automatically fired under computer control in a breach 306. Explosive gases are routed through a check valve 308 to a pressurized-gas tank 310. A single 4-gang solenoid valve 312 and 314 steers high pressure gas to and vents gases from pressurized water tanks 316 and 318. When one tank is being pressured, the other is being vented. A high-water float control inlet valve 320 automatically admits water to pressurized water tank 316 when the liquid level is below the operating range maximum and the other tank 318 is receiving gas pressure from explosive-gas tank 310 through 4-gang solenoid valve 312. Another high-water float control inlet valve 322 admits water to pressurized water tank 318 when its liquid level is below its operating range maximum and its gases are vented. Similarly, a low-water float control inlet valve 324 shuts off water from pressurized water tank 316 when the liquid level falls below the operating range minimum. Another low-water float control outlet valve 326 shuts off water from pressurized water tank 318 when its liquid level is below its operating range minimum. Pressure safety valves (PSV) 330, 331, and 332 release overpressures to protect the respective tanks from rupturing.

A water turbine 340 converts the hydraulic flow through it to a mechanical torque applied to a rotating driveshaft 342. A second water turbine 344 converts its hydraulic flow to additional mechanical torque that is also applied to rotating driveshaft 342. A liquid circuit 346 returns to pressurized water tank 316 through high-water float switch and valve 320. A DC electrical generator 348 converts the rotating mechanical torque to electrical power that is converted to AC by an inverter 350. Gears and pulleys in front of the generator may be used to adjust the speed and power input. Fill and drain valves are connected to the various tanks as appropriate. The system control signals may be supported on a computer network or conventional process control loops and can involve wireless connections.

A controller 352 operates the magazine 304 and breach 306, and valves 312 and 314 to coordinate their timing, such that gas pressure from the pressurized-gas tank 310 is alternately routed to each pressurized water tank 316 and 318 until the liquid inside is pushed out into the other. The inverter 350 provides load sensing signals to the controller 352. A throttle control 354 applied to control motors on valves 324 and 326 can be used to control the power output of turbines 340 and 344.

FIG. 4 represents a two-stage electrical generator embodiment of the present invention, and is referred to herein by the general reference numeral 400. Generator 400 uses explosive gases to pressurize water, and then uses two stages of pressurized hydraulics to spin electrical generators with hydraulic motors and turbines. A first Stage-1 uses explosive cartridges to produce hot gases that will pressurize a tank 402. Computer timing and valve control 404 steers the high pressure gas alternately to a first hydraulic pressure tank-A 406 and then to a second hydraulic pressure tank-B 408 according to their respective liquid levels. Water passing from the pressurized one of the tanks to the non-pressurized one is used to spin a hydraulic pump or water turbine 410. Vent gases recovered from hydraulic pressure tank-A 406 and tank-B 408 are captured by a second stage gas pressure tank 412.

The pressure loss in the gas pressures between the first Stage-1 and second Stage-2 is a function of the differential volumes of hydraulic pressure tank-A 406 and tank-B 408 as they cycle between their minimum and maximum water levels.

The second stage gas pressure tank 412 supplies gas to a computer timing and valve control 414 steers the high pressure gas alternately to a third hydraulic pressure tank-C 416 and then to a fourth hydraulic pressure tank-D 418 according to their respective liquid levels. Water passing from the pressurized one of these tanks to the non-pressurized one is used to spin a hydraulic pump or water turbine 420.

Both water turbines 410 and 420 can be geared to drive a single DC electric generator 422. The electrical power produced is temporarily stored in batteries 424, and that can smooth out any voltage variations that would otherwise result as the turbines are cycled between the hydraulic pressure tanks. An inverter 426 converts the DC power to AC power, and a transformer 428 is used to produce various commercial voltages, e.g., 110 VAC, 220-VAC, and 440-VAC at 50/60 Hertz.

Although the present invention has been described in terms of the presently preferred embodiments, it is to be understood that the disclosure is not to be interpreted as limiting. Various alterations and modifications will no doubt become apparent to those skilled in the art after having read the above disclosure. Accordingly, it is intended that the appended claims be interpreted as covering all alterations and modifications as fall within the "true" spirit and scope of the invention.

The invention claimed is:

1. A generator system, comprising:

a high pressure gas tank providing for the capture and confinement of gases generated by an explosive cartridge;

a magazine and breach connected to the high pressure gas tank, and providing for the operation of said explosive cartridge;

a pair of interconnected liquid tanks connected to receive gases routed from the high pressure gas tank, said liquid tanks containing a liquid and interconnected such that said liquids within flow in a circuit between the liquid tanks;

a high pressure turbine connected to be driven by said liquids flowing between the liquid tanks;

7

a low pressure gas tank connected to the pair of interconnected liquid tanks, and providing for the capture and confinement of gases vented from the interconnected liquid tanks;

a second pair of interconnected liquid tanks connected to receive gases routed from the low pressure gas tank, said liquid tanks containing a liquid and interconnected such that said liquids within flow in a circuit between the liquid tanks;

a low pressure turbine connected to be driven by said liquids flowing between the second pair of interconnected liquid tanks;

an electric generator connected to be driven by the high pressure and low pressure turbines and able to produce electrical power;

and a controller to operate valves and to coordinate the timing such that gas pressure from the pressurized gas tank is alternately routed to each liquid tank until the liquid inside is pushed out into the other;

wherein energy from said explosive cartridge is converted into electrical power.

2. The system of claim 1, further comprising: a high-water float switch and a low-water float switch disposed in each of the liquid tanks and connected to the controller; wherein the controller is enabled to maintain the liquid levels within each pair of interconnected liquid tanks over an operational range.

3. The system of claim 1, further comprising:

a liquid inlet valve providing a controlled input for each of the pair of interconnected liquid tanks that is connected in a circuit to receive liquids from the other liquid tank in the pair.

4. The system of claim 1, further comprising:

a liquid outlet valve providing a controlled output for each of the pair of interconnected liquid tanks that is connected in a circuit to transmit liquids to the other liquid tank in the pair.

5. The system of claim 1, further comprising:

a liquid inlet valve providing a controlled input for each of the pair of interconnected liquid tanks that is connected in a circuit to receive liquids from the other liquid tank in the pair;

a liquid outlet valve providing a controlled output for each of the pair of interconnected liquid tanks that is connected in a circuit to transmit liquids to the other liquid tank in the pair; and

a high-water float switch and a low-water float switch disposed in each of the liquid tanks;

wherein the liquid inlet valve and liquid outlet valve are controlled by the controller according to signals obtained from the high-water float switch and a low-water float switch.

6. A generator system, comprising:

a pressurized-gas tank providing for the capture and confinement of gases generated by an explosive cartridge;

a magazine and breach connected to the pressurized-gas tank, and providing for the operation of said explosive cartridge;

a pair of interconnected first and second liquid tanks connected to receive gases routed from the pressurized-gas tank, said liquid tanks containing a liquid and interconnected such that said liquids within flow in a circuit between the liquid tanks;

a first and a second liquid inlet valve providing a controlled input for each of the pair of interconnected liquid tanks that is connected in a circuit to receive liquids from the other liquid tank in the pair;

8

a first and a second liquid outlet valve providing a controlled output for each of the pair of interconnected liquid tanks that is connected in a circuit to transmit liquids to the other liquid tank in the pair;

a high-water float switch and a low-water float switch disposed in each of the first and second liquid tanks;

a first turbine connected to be driven by said liquids flowing from said first liquid tank to said second liquid tank;

a second turbine connected to be driven by said liquids flowing from said second liquid tank to said first liquid tank;

an electric generator connected to be driven by at least one of the first and second turbines and able to produce electrical power;

and a controller to operate the magazine and breach, and first and a second liquid inlet valve, and first and a second liquid outlet valves to coordinate their timing, such that gas pressure from the pressurized-gas tank is alternately routed to each liquid tank until the liquid inside is pushed out into the other, and according to signals obtained from the high-water float switch and low-water float switch;

wherein energy from said explosive cartridge is converted into electrical power.

7. A method of converting explosives energy into electrical power, comprising a computer program in software or firmware executed by a conventional microcomputer, with data inputs from sensors and switches digitized for processing, user inputs used to make process control decisions, and outputs to electro-mechanical solenoids to operate gas and hydraulic valves, comprising:

closing pressure tank and water tank vents, and closing water tank inlet valves;

raising a gas pressure in a pressure tank up to an operating level by firing explosive cartridges as needed;

checking a water level inside a water tank-A and if it's at its maximum operating level, beginning a hydraulic cycle, wherein a gas inlet valve-A is opened, a gas vent valve-A is closed, and a water outlet valve-A to an associated turbine-A is opened; wherein, a gas pressure let in can push water out through said outlet valve-A until the water level reaches a minimum, and said outlet valve-A is closed, said gas inlet valve-A is closed, and gas vent valve-A is opened, and water inlet valve-A is opened to receive water from water tank-B; and

checking a water level inside a water tank-B and if it's at its maximum operating level, beginning a hydraulic cycle, wherein a gas inlet valve-B is opened, a gas vent valve-B is closed, and a water outlet valve-B to an associated turbine-B is opened; wherein, a gas pressure let in can push water out through said outlet valve-B until the water level reaches a minimum, and said outlet valve-B is closed, said gas inlet valve-B is closed, and gas vent valve-B is opened, and water inlet valve-B is opened to receive water from water tank-A.

8. The method of claim 7, further comprising: if a user is not requesting a stop of operations, the process repeats in a loop.

9. The method of claim 7, further comprising: if a user is requesting a stop of operations, closing gas inlet pressure valves to water tank-A and tank-B, opening gas vents, and closing said water outlet valves to the turbines; wherein, residual gas pressures inside said pressurized tank is let down if another use cycle is not scheduled immediately.

* * * * *

ELECTRICAL GENERATION FROM EXPLOSIVES

Related Applications

This Application claims benefit of United States Provisional Patent Application, A UNIQUE METHOD OF GENERATING ELECTRICITY USING EXPLOSIVE SUBSTANCES AS A POWER SOURCE, serial number 61/100,915, filed 09/29/2008.

BACKGROUND OF THE INVENTION

Field of the Invention

The present invention relates to electrical power generation, and in particular to devices and methods for converting the gas pressure generated by explosives into electricity.

Description of the Prior Art

Useful electrical energy does not exist in nature and it must be converted from other available energy forms such as gasoline, diesel, coal, natural gas, geothermal, steam, hydro, solar, etc. Some of these energy sources are expensive, some are highly pollutant, some are difficult to convert efficiently, and some are not very portable as is needed in vehicles.

Explosives like gunpowder; Cordite, Ballistite, and Poudre-B smokeless powders; trinitrotoluene (TNT), Dynamite; nitroglycerin; Tovex and other water gel explosives; etc., release a lot of energy in a very rapid pulse. Explosives usually have less potential energy than petroleum fuels, but their high rate of energy release produces large blast pressures. TNT has a detonation velocity of 6,940 m/s compared to 1,680 m/s for the velocity of 6,940 m/s compared to 1,680 m/s for the detonation of a pentane-air mixture, and the 0.34-m/s stoichiometric flame speed of gasoline combustion in air. Explosives are classified as "low" or "high" explosives according to their rates of decomposition. Gunpowder is a low explosive, while TNT is a high explosive. Low explosives burn rapidly or deflagrate, while high explosives detonate.

The energy released includes high levels of heat, light, and gas pressure. These are all quickly dissipated if not captured or otherwise contained. For example, at 15°C the volume of gas produced by the explosive decomposition of one mole of nitroglycerin, becomes, V = (23.64 liter/mol)(7.25 mol) = 171.4 liters. The molar volume of an ideal gas at 15°C is about 23.64 liters. The potential of an explosive is the total work that can be performed by the gas generated by the explosion. If uncontained, it expands adiabatically from its original volume until its pressure is reduced to atmospheric pressure and its temperature to ambient.

In the nitroglycerin reaction, $C_3H_5(NO_3)3 \rightarrow 3CO_2 + 2.5H_2O + 1.5N_2 + 0.25O_2$, the products are carbon dioxide, water, nitrogen, oxygen, and heat. Therefore, a relatively small solid or liquid volume is converted into a very large volume of relatively benign gases. Nitroglycerin explosions are relatively clean, compared to TNT which is poisonous and produces a lot of carbon soot in its reaction.

Firearms and artillery use the gas pressure generated by the detonation of smokeless powder to accelerate bullets and projectiles to very high muzzle velocities on the order of 2,000+ feet per second. Sticks of explosives are detonated in holes drilled into geologic deposits to fracture the ores and make removing the material as easy as scooping up the pieces.

What is needed is a device and a method to convert explosive energy into a more useful form of electrical energy as used in homes and industry; detonated in holes drilled into geologic deposits to fracture the ores and make removing the material as easy as scooping up the pieces.

SUMMARY
OF THE
INVENTION

Briefly, an electrical generator embodiment of the present invention converts the high blast pressures of explosives into useful electricity by capturing the explosive gases and using the high gas pressures to alternately push water hydraulically between two tanks and through water turbines connected to DC electric generator. Water expelled through a water turbine from one tank is used to fill the other tank. Batteries can be used to store the electrical energy generated, and inverters followed by transformers convert the DC electric from the turbine-generators to 110-VAC, 220-VAC, and 440-VAC. A microcomputer controller connected to various sensors and solenoid valves coordinates the detonation of the explosives, tank pressures, venting, valving, and load control.

These and other objects and advantages of the present invention will no doubt become obvious to those of ordinary skill in the art after having read the following detailed description of the preferred embodiments which are illustrated in the various drawing figures.

IN THE DRAWINGS

Fig. 1   is a simplified functional block diagram of a single-stage electrical generator embodiment of the present invention that cycles pressurized water between two tanks and through two sets of water turbines;

Fig. 2   is a flowchart diagram of an electrical generator method embodiment of the present invention to cycle pressurized water between two tanks and two water turbines, as in Fig. 1;

Fig. 3   is a functional block diagram of a single-stage electrical generator embodiment of the present invention that cycles pressurized water between two tanks and through two water turbines like that of Fig. 1, but that reduces duplication of the DC generators and inverters, and some of the valving; and

Fig. 4   is a simplified functional block diagram of a two-stage electrical generator embodiment of the present invention that uses explosive gases to pressurize water, and then uses pressurized hydraulics to spin electrical generators with hydraulic motors and turbines.

COMPONENT NUMBERING SYSTEM FOR FIGURES 1 THRU 4

Component Numbering for Fig. #1

100 Single Stage Electrical Invention

102 Cartridges

104 Magazine

106 Breach

108 Check Valve

110 Explosive Gases Pressure Vessel

112 Pressure Safety Valve (Interfaced with Computer by Wireless)

114 Pressure Sensing Valve (Interfaced with the Computer by Wireless)

120 Microcomputer Controller

H High Float Switch Sensor

L Low Float Switch Sensor

122 Pressurized Gas Outlet Valve for 110; Inlet Valve for 126; Equipped with Wireless Interface with 120

124 Pressurized Gas Outlet Valve for 110; Inlet Valve 128; Equipped with Wireless Interface with 120

126 Pressurized Water Tank- A

128 Pressurized Water Tank- B

130 Water Tank Outlet Valve from 126; Inlet Valve for 134: Equipped with Wireless Interface with 120

134 Water Turbine – A

136 Water Turbine – B

138 Water Inlet Valve to 126: Equipped with Wireless Interface with 120

140 Water Outlet Valve from 134; Inlet Valve for 128; Equipped with Wireless Interface with 120

142 & 144 Low and High Water Switches (L,H) for Water Tank-A; Equipped with Wireless Interface

146 & 148 Low and High Water Switches (L,H) for 128; Equipped with Wireless Interface with 120 Direct Current Generator

152 Inverter for 150

154 Direct Current Generator

156 Inverter for 154`

160 Vent Valve for 126; Equipped with Wireless Interface with 120

162 Vent Valve for 128; Equipped with Wireless Interface with 120

170 Nozzle Control Valve for 134

172 Nozzle Control Valve for 136

Component Numbering System for
Fig. #2
Component Numbering Block Diagrams

200 Start Up (Yes/No)

204 Vent Closure and Valve Closer for 126 and 128

206 Conditions for Loading Cartridges

208 Operational Instructions concerning Activation of Valves with 126 Operational

210 Operational Instructions concerning Activation of Valves with 128 Operational

212 Instructions for Stopping Operation

214 Shut Down of the System

Component Numbering System for
Fig. #3
Component Numbering

300 Single Stage Electrical Invention

302 Cartridges

304 Magazine

306 Breach

308 Check Valve

310 Explosive Gases Pressure Vessel

312 Gang Pressure Outlet Double Valve

314 Gang Pressure Outlet Double Valve to Secondary Explosive Gases Tank

316 Pressurized Water Tank-C

318 Pressurized Water Tank-D

320 High Water Switch in Tank-C equipped with Wireless Internet to 352

322 High Water Switch in Tank-D equipped with Wireless Interconnect to 352

324 Low Water Switch in Tank-C equipped with Wireless Interconnect to 352 & Throttle

326 Low Water Switch in Tank-D equipped with Wireless Interconnect to 352 & Throttle #2

331 Pressure Safety Vent Valve for Tank-C equipped with Wireless Interconnect to 352

332 Pressure Safety Vent Valve for Tank-D equipped with Wireless Interconnect to 352

340 Water Turbine #1

344 Water Turbine #2

346 Liquid Circuit from 344 to 316

Component Numbering System for
Fig. #4
Component Numbering

400 System Operation shown in Block Form

402 Exploded gas in Primary Pressure Tank

404 Timing and Valving Operation Control

406 Hydraulic Pressure Tank-A

408 Hydraulic Pressure Tank-B

410 Primary Hydraulic Turbine

412 Second Stage Pressurized Gas Tank

414 Secondary Timing and Valving Operational Control

416 Secondary Hydraulic Tank-C

418 Secondary Hydraulic Tank-D

420 Secondary Hydraulic Turbine

422 Generator Jointly for 410 and 420

424 Battery Pack

426 Inverter

428 Multi-tap Transformer

DETAILED DESCRIPTION OF THE PREFERRED EMBODIMENT

In **Figs. 1-4** and the following text, some of the more conventional and routine elements commonly used with gas and hydraulic valves, pressure tanks, plumbing, and process control systems are not shown or described. For example, inspection ports and drains for water tanks, safety relief valves, check valves, nozzles for turbines, gearboxes and pulleys, wireless interfaces, wiring, etc. The components like these that should be used are engineering choices and are routinely stocked and installed by technicians. The critical and unusual combinations and their interrelationships are described here in detail.

Fig. 1 represents a single-stage electrical generator system embodiment of the present invention, and is referred to herein by the general reference numeral **100**. Generator system **100** produces electrical power suitable for homes, businesses, industry, and the utility grid from the explosive energy captured from cartridges **102** loaded in a magazine **104** and fired in a breach **106**. Cartridges **102** should include low explosives that burn clean and soot-free, and the chemical reactions should not produce any dangerous gases or byproducts. For example, nitroglycerin reactions only produce carbon dioxide, water, nitrogen, oxygen, and heat. The heat actually helps increase the gas pressures up to operating levels and should not be wasted or exhausted until the maximum in work has been extracted.

The heated gaseous explosive products are passed through a check valve **108** to a pressurized-gas tank **110.** A pressure safety valve (**PSV**) **112** provides relief if the internal pressures exceed a safe maximum. A pressure sensor (**P**) **114** measures the tank pressures for a microcomputer controller **120**. In some installations the simultaneously pressurized, the pressure applied to them is alternated by gas pressure inlet valves **122** and **124** under control of microcomputer controller **120**. What's important to the timing is the water levels inside the tanks, there are minimum and maximum operating levels that must be respected. Water inside one tank needs to flow out into the other tank through a water turbine, and the water cannot flow if the receiving water tank is pressurized at the same time.

In **Fig. 1**, an outlet valve-A **130** is opened to pass pressured liquid water (**L**) to a water turbine-A **134**. Similarly, an outlet valve-B **132** is opened to pass pressured liquid water (**L**) to a water turbine-B **136**. The liquid water returns from water turbine-B **136** through an inlet valve-A **138** back to water tank-A **126**. Liquid **138** back to water tank-A **126**. Liquid water from water turbine-A **134** passes on through to inlet valve-B **140** to water tank-B **128**.

The minimum and maximum operating levels of water that circulate between water tank-A **126** and water tank-B **128** are set by float switches (**L, H**) **142** and **144** for water tank-A **126**, and by float switches (**L, H**) **146** and **148** for water tank-B **128**. These float switches are connected to microcomputer controller **120**, and the readings are used to determine when to open and close outlet valve-A **130**, outlet valve-B **132**, inlet valve-A **138**, and inlet valve-B **140**. The float switch connection could be done wirelessly, and a local loop could be included to automatically close, for example, water outlet valve-A **130** when minimum level float switch **142** senses low water.

Each water tank-A **126** and water tank-B **128** should be equipped with a water to add make up water, and to drain water completely, e.g., during maintenance.

In **Fig. 1**, water turbine-A **134** is mechanically connected by a rotating shaft to drive a DC electrical generator **150.** The DC electrical power produced could be stored in batteries, and it is converted to AC electrical power by an inverter **152.** Similarly, water turbine-B **136** is mechanically connected by a rotating shaft to drive another DC electrical generator **154.** The DC electrical power produced could be stored in the same batteries, and it can also be independently converted to AC electrical it can also be independently converted to AC electrical power by an inverter **156.** The voltage outputs of inverters 152 and 156 can be stepped-up or stepped-down by conventional transformers as needed, e.g., to 110-VAC, 220-VAC, and 440-VAC.

Microcomputer controller **120** is connected to sense the electrical loads placed on inverters **152** and **156**, and uses the information to control how much pressurized water is needed to be passed through water turbine-A **134** and water turbine-B **136** to keep the overall operation in balance.

Once the pressurized gas inside the water tanks has done its job pushing out the water down to its minimum operating level, the residual pressurized gas can be vented out. For water tank-A **126**, a vent valve-A **160** is used, and for water tank-B **128**, a vent valve-A **162** is used. The residual gas pressures can be high enough to do useful work in a second stage generator. But any back pressure caused by the use of later stages can reduce the efficiency of the earlier stages by reducing the differential pressures between the pressurized tank and the vented one.

In operation, falling water levels inside the water tanks can be used by the minimum-level float switches **142** and **146** to trigger closed the associated water outlet valves **130** and **132**. This, in turn can be used to trigger closed the gas pressure inlet valves **122** and **124,** and to trigger open the gas pressure vent valves **160** and **162.** Similarly, the maximum-level float switches **144** and **148** can be used to trigger closed the water inlet valves **138** and **140.**

Pressurized water tank-A **126** and pressurized water tank-B **128** would normally be equipped with various conventional items not shown in **Fig. 1.** For example, inspection ports, drain valves, pressure gauges, pressure safety valves to release excess pressure, and a water make-up input to replace lost water.

Microcomputer controller **120** can increase and decrease the torque outputs of water turbine-A **134** and water turbine-B **136** by sending modulation controls to nozzle controls **170** and **172.** Alternatively, water outlet valve-A **130** and water outlet valve-B **132** could be continuously adjustable, instead of simple fully open, fully closed solenoid types. Such torque modulation would be necessary in some applications to balance the power being generated with the loads applied. In such case, inverters **152** and **156** would also be required to provide load measurements to microcomputer controller **120.**

Fig. 2 represents an electrical generator method embodiment of the present invention to cycle pressurized water between two tanks and two water turbines, as in **Fig. 1**. Such method is referred to herein by the general reference numeral **200.** Method **200** is implemented as a computer program in software or firmware executed by a conventional microcomputer, e.g., microcomputer controller **120 (fig. 1).** Data inputs from sensors and switches are digitized for processing, user inputs are used to make process control decisions, and outputs to electro-mechanical solenoids are used to operate gas and hydraulic valves.

Method **200** includes three phases of operation: (**1**) startup, (**2**) power generation, and (**3**) shutdown. During startup, the operational pressures and valve settings must be initialized. During power generation, the gas pressure generated by the explosive cartridges must be switched between the two water tanks according to the respective water levels inside each. The amount of water forced water levels inside each. The amount of water forced between the water tanks and through the water turbines must be balanced with the electrical loads being placed on the system. During shutdown, the cycling must be stopped and the pressures relieved by opening the various vents.

Specifically, method **200** includes a step **202** for checking to see if the user wants to begin operation. If so, a step **204** closes the pressure tank and water tanks vents, and closes the inlet valves to the water tanks. A step **206** gets the gas pressure in the pressure tank up to operating levels by firing explosive cartridges as needed. A step 208 checks the water level inside water tank-A and if it's at its maximum operating level, a hydraulic cycle can begin. The gas inlet valve-A is opened, the gas vent valve-A is closed, and the water outlet valve-A to the associated turbine-A is opened. The gas pressure let in will push the water out through the outlet valve-A. When the water level reaches minimum, the outlet valve-A is closed. The gas inlet valve-A is closed, and the gas vent valve-A is opened. The water inlet valve-A is opened to receive water from water tank-B.

A step **210** checks the water level inside water tank-B and if it's at its maximum operating level, a hydraulic cycle can begin. The gas inlet valve-B is opened, the gas vent valve-B is closed, and the water outlet valve-B to the associated turbine-B is opened. The gas pressure let in will push the water out through the outlet valve-B. When the water level in water tank-B reaches minimum, the outlet valve-B is closed. The gas inlet valve-B is closed, and the gas vent valve-B is opened. The water inlet valve-B is opened to receive water from water tank-A.

If the user is requesting a stop of operations, a step **212** passes control to a step **214.** Otherwise, the process repeats in a loop back to step **206.** Step **214** closes the gas inlet pressure valves to water tank-A and tank-B, opens the vents, and closes the water outlet valves to the turbines. Residual gas pressures inside the pressurized tank may be let down if another use cycle is not expected immediately.

Fig. 3 illustrates a single-stage system **300** that eliminates some of the duplication of the major components appearing in **Fig. 1.** System **300** assumes that when the water level in a water tank is below minimum, e.g., as detected by a low-water float switch, the water outlet valve should be closed. Similarly, when the water level in a water tank is above maximum, e.g., as detected by a high-water float switch, the water inlet valve should be closed. The gas inlet valve to a water tank can only be open if the gas vent is closed. The gas inlet valve to the water tank must be closed if the gas vent is open. The mechanisms implemented to enforce such logic can be built with relay logic, software, IC logic gates, and mechanical interlocks.

System **300** is powered by explosive cartridges **302** that are loaded in a magazine **304** and automatically fired under computer control in a breach **306**. Explosive gases are routed through a check valve **308** to a pressurized-gas tank **310**. A single 4-gang solenoid valve **312** and **314** steers high pressure gas to and vents gases from pressurized water tanks **316** and **318**. When one tank is being pressured, the other is being vented. A high-water float control inlet valve **320** automatically admits water to pressurized water tank **316** when the liquid level is below the operating range maximum and the other tank **318** is receiving gas pressure from explosive-gas tank **310** through 4-gang solenoid valve **312**. Another high-water float control inlet valve **322** admits water to pressurized water tank **318** when its liquid level is below its operating range maximum and its gases are vented. Similarly, a low-water float control inlet valve **324** shuts off water from pressurized water tank **316** when the liquid level falls below the operating range minimum. Another low-water float control outlet valve

326 shuts off water from pressurized water tank **318** when its liquid level is below its operating range minimum. Pressure safety valves **(PSV) 330, 331,** and **332** release overpressures to protect the respective tanks from rupturing.

A water turbine **340** converts the hydraulic flow through it to a mechanical torque applied to a rotating driveshaft **342**. A second water turbine **344** converts its hydraulic flow to additional mechanical torque that is also applied to rotating driveshaft **342**. A liquid circuit **346** returns to pressurized water tank **316** through high-water float switch and valve **320**. A DC electrical generator **348** converts the rotating mechanical torque to electrical power that is converted to AC by an inverter **350**. Gears and pulleys in front of the generator may be used to adjust the speed and power input. Fill and drain valves are connected to the various tanks as appropriate. The system control signals may be supported on a computer network or conventional process control loops and can involve wireless connections.

A controller **352** operates the magazine **304** and breach **306**, and valves **312** and **314** to coordinate their timing, such that gas pressure from the pressurized-gas tank **310** is alternately routed to each pressurized water tank **316** and **318** until the liquid inside is pushed out into the other. The inverter **350** provides load sensing signals to the controller **352**. A throttle control **354** applied to control motors on valves **324** and **326** can be used to control the power output of turbines **340** and **344.**

Fig. 4 represents a two-stage electrical generator embodiment of the present invention, and is referred to herein by the general reference numeral **400**. Generator 400 uses explosive gases to pressurize water, and then uses two stages of pressurized hydraulics to spin electrical generators with hydraulic motors and turbines. A first Stage-1 uses explosive cartridges to produce hot gases that will pressurize a tank **402**. Computer timing and valve **318** until the liquid inside is pushed out into the other. The inverter **350** provides load sensing signals to the controller **352**. A throttle control **354** applied to control motors on valves **324** and **326** can be used to control the power output of turbines **340** and **344**.

Fig. 4 represents a two-stage electrical generator embodiment of the present invention, and is referred to herein by the general reference numeral **400**. Generator **400** uses explosive gases to pressurize water, and then uses two stages of pressurized hydraulics to spin electrical generators with hydraulic motors and turbines. A first Stage-1 uses explosive cartridges to produce hot gases that will pressurize a tank **402**. Computer timing and valve control **404** steers the high pressure gas alternately to a first hydraulic pressure tank-A **406** and then to a second hydraulic pressure tank-B **408** according to their respective liquid levels. Water passing from the pressurized one of the tanks to the non-pressurized one is used to spin a hydraulic pump or water turbine **410.** Vent gases recovered from hydraulic pressure tank-A **406** and tank-B **408** are captured by a second stage gas pressure tank **412.**

The pressure loss in the gas pressures between the first Stage-1 and second Stage-2 is a function of the differential volumes of hydraulic pressure tank-A **406** and tank-B **408** as they cycle between their minimum and maximum water levels.

The second stage gas pressure tank **412** supplies gas to a computer timing and valve control **414** steers the high pressure gas alternately to a third hydraulic pressure tank-C **416** and then to a fourth hydraulic pressure tank-D **418** according to their respective liquid levels. Water passing from the pressurized one of these tanks to the non-pressurized one is used to spin a hydraulic pump or water pressurized from one is used to spin a hydraulic pump or water turbine **420.**

Both water turbines **410** and **420** can be geared to drive a single DC electric generator **422.** The electrical power produced is temporarily stored in batteries **424,** and that can smooth out any voltage variations that would other wise result as the turbines are cycled between the hydraulic pressure tanks. An inverter **426** converts the DC power to AC power, and a transformer **428** is used to produce various commercial voltages, e.g., 110 VAC, 220-VAC, and 440-VAC at 50/60 Hertz.

Although the present invention has been described in terms of the presently preferred embodiments, it is to be understood that the disclosure is not to be interpreted as limiting. Various alterations and modifications will no doubt become apparent to those skilled in the art after having read the above disclosure. Accordingly, it is intended that the appended claims be interpreted as covering all alterations and modifications as fall within the "true" spirit and scope of the invention.

CLAIMS

1. A generator system, comprising:

    a pressurized-gas tank providing for the capture and confinement of gases generated by an explosive cartridge;

    a pair of interconnected liquid tanks for pressurizing liquids inside with gases routed from the pressurized-gas tank, said liquid tanks interconnected such that said liquids within flow in a circuit between the liquid tanks;

    a turbine connected to be driven by said liquids flowing between the liquid tanks;

    a electric generator connected to be driven by the turbine and able to produce electrical power; and

    a controller to operate valves and to coordinate the timing such that gas pressure from the pressurized-gas tank is alternately routed to each liquid tank until the liquid inside is pushed out into the other;

wherein, energy from said explosive cartridge is converted into electrical power.

2. The system of Claim 1, further comprising:

a magazine and breach connected to the pressurized-gas tank, and providing for the operation of said explosive cartridge.

3. The system of Claim 1, further comprising:

a high-water float switch and a low-water float switch disposed in each of the liquid tanks and connected to the controller;

wherein the controller is enabled to maintain the liquid levels within the pair of interconnected liquid tanks over an operational range.

4. The system of Claim 1, further comprising:

a liquid inlet valve providing a controlled input for each of the pair of interconnected liquid tanks that is connected in a circuit to receive liquids from the other liquid tank in the pair.

5.   The system of Claim 1, further comprising:

a liquid outlet valve providing a controlled output for each of the pair of interconnected liquid tanks that is connected in a circuit to transmit liquids to the other liquid tank in the pair.

6.   The system of Claim 1, further comprising:

a liquid inlet valve providing a controlled input for each of the pair of interconnected liquid tanks that is connected in a circuit to receive liquids from the other liquid tank in the pair;

a liquid outlet valve providing a controlled output for each of the pair of interconnected liquid tanks that is connected in a circuit to transmit liquids to the other liquid tank in the pair; and

a high-water float switch and a low-water float switch disposed in each of the liquid tanks;

wherein the liquid inlet valve and liquid outlet valve are controlled by the controller according to signals obtained from the high-water float switch and a low-water float switch.

7.   A generator system, comprising:

a pressurized-gas tank providing for the capture and confinement of gases generated by an explosive cartridge;

a magazine and breach connected to the pressurized-gas tank, and providing for the operation of said explosive cartridge;

a pair of interconnected first and second liquid tanks for pressurizing liquids inside with gases routed from the pressurized-gas tank, said liquid tanks interconnected such that said liquids within flow in a circuit between the liquid tanks;

a first and a second liquid inlet valve providing a controlled input for each of the pair of interconnected liquid tanks that is connected in a circuit to receive liquids from the other liquid tank in the pair;

a first and a second liquid outlet valve providing a controlled output for each of the pair of interconnected liquid tanks that is connected in a circuit to transmit liquids to the other liquid tank in the pair; and

a high-water float switch and a low-water float switch disposed in each of the first and second liquid tanks;

a first turbine connected to be driven by said liquids

flowing from said first liquid tank to said second liquid tank;

a second turbine connected to be driven by said liquids flowing from said second liquid tank to said first liquid tank;

an electric generator connected to be driven by at least one of the first and second turbines and able to produce electrical power; and

a controller to operate the magazine and breach, and first and a second liquid inlet valve, and first and a second liquid outlet valves to coordinate their timing, such that gas pressure from the pressurized-gas tank is alternately routed to each liquid tank until the liquid inside is pushed out into the other, and according to signals obtained from the high-water float switch and low-water float switch;

wherein, energy from said explosive cartridge is converted into electrical power.

8.   A method of converting explosives energy into electrical power, comprising a computer program in software or firmware executed by a conventional microcomputer, with data inputs from sensors and switches digitized for processing, user inputs used to make process control

decisions, and outputs to electro-mechanical solenoids to operate gas and hydraulic valves, comprising:

three phases of operation including startup, power generation, and shutdown;

wherein, during startup, operational pressures and valve settings are initialized;

wherein, during power generation, gas pressure generated by explosive cartridges is switched between two water tanks according to respective water levels inside each, and an amount of water forced between said water tanks and through the water turbines is balanced with any electrical loads being placed on the system;

wherein, during shutdown, any cycling is stopped and any tank pressures are relieved by venting.

9.   The method of Claim 8, further comprising:

checking to see if a user wants to begin operation, and if so, closing pressure tank and water tank vents, and closing water tank inlet valves.

10.   The method of Claim 8, further comprising:

raising a gas pressure in a pressure tank up to an operating level by firing explosive cartridges as needed.

11.   The method of Claim 8, further comprising:

checking a water level inside a water tank-A and if it's at its maximum operating level, beginning a hydraulic cycle, wherein a gas inlet valve-A is opened, a gas vent valve-A is closed, and a water outlet valve-A to an associated turbine-A is opened;

wherein, a gas pressure let in can push water out through said outlet valve-A until the water level reaches a minimum, and said outlet valve-A is closed, said gas inlet valve-A is closed, and gas vent valve-A is opened, and water inlet valve-A is opened to receive water from water tank-B;

checking a water level inside a water tank-B and if it's at its maximum operating level, beginning a hydraulic cycle, wherein a gas inlet valve-B is opened, a gas vent valve-B is closed, and a water outlet valve-B to an associated turbine-B is opened; and

wherein, a gas pressure let in can push water out through said outlet valve-B until the water level minimum, and said outlet valve-B is closed, said gas inlet valve-B is closed, and gas vent valve-B is opened, and water inlet valve-B is opened to receive water from water tank-A.

12.　The method of Claim 10, further comprising:

if a user is not requesting a stop of operations, the process repeats in a loop.

13.　The method of Claim 10, further comprising:

if a user is requesting a stop of operations, closing gas inlet pressure valves to water tank-A and tank-B, opening gas vents, and closing said water outlet valves to the turbines;

wherein, residual gas pressures inside said pressurized tank is let down if another use cycle is not scheduled immediately.

ABSTRACT
OF THE
DISCLOSURE

An electrical generator converts the high blast pressures of explosives into useful electricity by capturing the explosive gases and using the high gas pressures to alternately push water hydraulically between two tanks and through water turbines connected to DC electric generators. Water expelled through a water turbine from one tank is used to fill the other tank. Batteries can be used to store the electrical energy generated, and inverters followed by transformers convert the DC electric from the turbine-generators to 110-VAC, 220-VAC, and 440-VAC. A microcomputer controller connected to various sensors and solenoid valves coordinates the timing and routing of the detonation of explosives, tank pressures, venting, valving, and load control.

TITLE OF THE INVENTION: ELECTRICAL GENERATION FROM EXPLOSIONS

Additional Claims are hereby being submitted in this Document.

Added Claims:

Claim No. 13

1. Said Invention is to be free standing without any reliance on external items except as listed below:

The battery pack **Fig, 4; # 424** is to be fully charged up initially from an external power source. Subsequent operation shall be performed using internal power (Battery Pack (# **424**) and Generator (#**422**).

Tank No. 1 (**#126**) is to be filled up with fluid up to the Upper Float (**#144**) in the fully upright position.

Tank No. 2 (#**128**) is to be filled up with fluid up to the Lower Float (#**146**) fully upright position.

Make-up fluid shall be provided as needed.

In cold climates where freeze up of the system exists an Anti Freeze solution shall be added to the fluid (water in this case). A mixture of ethanol glycol and water can freeze at -45 Fahrenheit.

Claim No. 14

This Invention consists of one Pressure Vessel (#**110**); two pressurized water tanks (#**126** & #**128**); a battery pack (#**424**); a micro-controller (#**120**); a generator (**Fig, #1; #150**); two turbines (#**134** & #**136**); a breach (#**106**); and a magazine for explosive cartridges (#**104**).

Claim No.15

All of the above components shall be assembled on a metal base such that the **Center of Gravity** is midway laterally and longitudinally and midway vertically.

Design Requirements for Use on a Space Probe or Space Craft

Note: The Boost Vectors must all pass through the Center of Gravity of the entire Space Probe during the Boost Phase otherwise tumbling of the Vehicle may ensue. The Boost Phase occurs during the Launch to Orbit Phase. This is the High Acceleration Phase and buffeting occurs until the Vehicle passes through and out of the Earth's atmosphere.

Claim #16

The metal base upon which the above components are mounted shall have openings on each side to accommodate the blades of a fork-lift for transportability.

The base shall be furnished with metal wheels for ease in moving the assembly about. On each corner of the base there shall be casters wheels on one end and fore and aft wheels on each corner at the rear. The wheels shall have wide metal rims.

Each wheel shall have a locking device.

Hoisting fixtures shall be furnished on the upper corner of the overall frame for lifting by a crane or other lifting devices for loading aboard ship or otherwise.

A sub-frame shall be furnished for mounting the assembly on for transporting same. The sub-frame shall support the assembly by raising the wheels off of any surface.

Claim #17

There shall be two (2) versions of the Assembly.

There shall be Version #1 and Version #2.

Version #1: This version shall be built out of light metal but is to be designed for use in the Earths Gravitational Field. Weigh in this Version is not considered critical.

Version #2: This Version shall be built with weigh reduction considered to be critical without jeopardizing the structural Integrity of the Assembly's capability to Operate as Intended.

This Version is intended to be designed to function in any area where a Gravitational Field exists even though it may not be as strong as the Earth's Field.

Version #2 is to be designed for installation in a subterranean air tight structure on the moon. It is understood that such a chamber or room is to have climatic control of temperature as well as furnishing an atmosphere for life support. The Gravity on the moon is approximately 1/6 that on earth. A sufficient supply of explosive cartridges shall also be included to last the life of the Mission.

Version #2 is to be built with very light weight to reduce the overall weight of the Launch Vehicle. It shall be transported by being suspended from the top; hanging like a basket. The Assembly might collapse if supported on the base during the Boost Phase (high acceleration).

In preparation for a flight in space the Pressure Vessel (**#110**) shall be charged up to its maximum pressure prior to Lift-Off.

Gases under high pressure from the Pressure Vessel shall be used to power the Reaction Control Nozzle System.

Burst of Gases shall be emitted from the proper orientation nozzle(s) to assist in maintaining the proper orientation of the Space Craft.

If this Invention is also included in an un-manned Space Probe the High Pressure Gases shall be utilized in at least two (2) ways.

Usage #1: The High Pressure Gases shall assist in maintaining the proper orientation of the Space Craft by providing short burst of gases from the proper nozzle or nozzles.

In addition, burst of gases from a separate nozzle or set of nozzles that are oriented in such as way as to accelerate the Space Probe shall be used for such a purpose.

A Separate set of nozzles that are orientated to slow the Space Probe shall also use burst of gases under pressure to accomplish this purpose also.

This Invention can be used to provide electricity for the Space

Probe if the Assembly Base is oriented at right angles to the path of the Space Probe and is mounted with the Base at the end opposite to the direction of acceleration. The inertia of the Assembly results in a force similar to gravity during an acceleration phase.

If it is desired to slow the Space Probe (Like a negative gravity the Invention Assembly must be situated at the opposite end with the base in a direction opposite to the slow down deceleration vector.

This operation requires that the Space Probe be reoriented. This kind of situation must exist if the Invention is to provide electricity for usage during this part of the mission.

It is assumed that the Space Probe has been provided with Climatic Control in the Interior; in the Equipment Bay.

Claim #18
In the event that a Lunar Subterranean Base has been established then perhaps the ingredients for making explosives would be included in the Space Craft. A device of some sort would be required at the job site to manufacture the explosive substances.

SUMMARY

Advantages of the Invention

1. Limited emission of Carbon Dioxide

2. Reliability

3. Portability

4. Safe in Operation

5. Convenience

6. Ease of Operation

7. Adaptability

8. Intermittent Operation as concerns the detonation of Explosive Substances. It does not involve repetitive explosions.

9. Simplicity of Design

TESTING OF THE INVENTION

To confirm the advisability of the practicality of placing it in production the following is required:

Two (2) moderate sized butane tanks

One (1) Off the shelf hydro-electric generator

Metal piping

Various manually operated valves and switching

A supply of water

A Pressure Valve

A Vent Valve

A Fill opening with a screw on cap

A Drain Valve

A Volt Meter

An Amp Meter

A Watt Meter

A Load such as an Electric light etc.

Services for drilling holes in the tanks; welding pipes

An Air Compressor capable of producing high pressure air on the order of 400 psi. This substitutes for the detonation of the Explosive Substance.

Services of an Engineer verifying that the butane tanks can safely contain the high pressure air

End

X